ATMOSPHERIC POLLUTION BY INDUSTRIAL WASTE

THE CHANDLER LECTURE
1923

COLUMBIA UNIVERSITY PRESS
COLUMBIA UNIVERSITY
NEW YORK

SALES AGENTS
LONDON
HUMPHREY MILFORD
AMEN CORNER, E. C.

SHANGHAI
EDWARD EVANS AND SONS, LTD.
30 N. SZECHUEN ROAD

ATMOSPHERIC POLLUTION BY INDUSTRIAL WASTES

BY
ROBERT E. SWAIN
Leland Stanford University, Stanford University, Calif.

New York
COLUMBIA UNIVERSITY PRESS
1923

All rights reserved

Copyright, 1923
By COLUMBIA UNIVERSITY PRESS

Printed from type, June 1923

Printed by Eschenbach Printing Co.
Easton, Pa., U. S. A.

Robert E. Swain

Atmospheric Pollution by Industrial Wastes

By Robert E. Swain

THE GREATEST single waste product in industry is a gas, carbon dioxide, which is usually discharged as it is produced, directly into the air. The annual production of this substance reaches figures, in terms of the usual units of volume or weight in which we measure gases, which are of the order of those in which we express astronomical dimensions. If all the coal consumed annually in this country were completely burned, there would be produced approximately nine hundred thousand billion cubic meters, or one billion eight hundred million metric tons of this gas. The combustion of petroleum would add two hundred million metric tons, and of natural gas, ninety million metric tons, while the burning or decay of wood, and of plant products and tissues, would add an indeterminable but enormous total to these figures. Our fellow human beings in this country alone exhale an annual total of about forty-five million metric tons of this gaseous product of respiration. And we are overlooking here a great many other sources of carbon dioxide, both large and small. But it is a remarkable fact that, rapidly dispersed as it is into the great atmospheric ocean about us, this gas is present in the strikingly small and uniform amount of three parts by volume of carbon dioxide to ten thousand parts of air, or three hundred parts per million parts of air. Except in the air over large cities on still days, where values as high as six parts per ten thousand parts of air have been found, this proportion varies but little over land and sea, at high altitudes and at low. There is here an impressive illustration of the ability of the atmosphere to dissipate rapidly to low concentrations gaseous substances which are discharged into it. In spite of this, however, injurious concentrations of certain other gaseous waste products may often arise in industrial regions.

Carbon dioxide is not only an enormous waste product; it is the most important one in its relation to life on this universe, for it is absorbed by the green leaves of plants and by them converted, by a process called photosynthesis, to substances which are foods for animals. Agriculture is but the means of stimulating this process through the cultivation, under the most favorable conditions, of those plants which are found to be of most service to man.

COAL AS SOURCE OF POLLUTION

Coal, on one hand, and sulfur-containing ores, on the other, are easily foremost as ultimate sources of the substances which concern us most in atmospheric pollution. By the average citizen coal is regarded simply as a source of heat energy, but if he is a resident of a large industrial area in which much soft coal is burned, he knows that it is the chief source of the hazy, soot-laden atmospheres which are all too prevalent as symbols of waste and enemies of civic cleanliness. During the Great War the Fuel Administration estimated that between twenty-five and thirty million tons of coal could be saved in this country every year by the industries alone through more efficient methods of firing. Another estimate places at 20 per cent the annual loss in soot and other products, solid and gaseous, of incomplete combustion of coal. This represents much more of an economic loss than simply one of energy, for it clouds the atmosphere over our great cities, does extensive damage to buildings, furnishings, and merchandise, and leads to widespread personal loss and discomfort. In many industrial centers the sooty, tarry deposits run as high as six hundred tons per square mile per annum, while results as high as four hundred tons are not uncommon in the winter months in cities of moderate size. It has been claimed, particularly by English investigators, that the unflourishing condition of trees and other vegetation in large cities and industrial centers with smoky atmospheres is due largely to the deposition on the leaf parts of the tarry products of the incomplete combustion of coal. The conifers as a class have suffered most. Since they are evergreen, their leaves are exposed to the deposition of sooty substances during the entire year, and particularly during the winter season when the deposition is greatest.

THE SULFUR DIOXIDE PROBLEM

Another phase of the smoke problem which I want particularly to discuss is industrial smoke arising from metallurgical operations. This question has come up out of the West rather than of the East, and has left in its train many years of bitter litigation in which agrarian and industrial interests have clashed. And in the West the chief contributions to the solution of the question have been made.

Among the waste products of metallurgical operations the conspicuous offender is sulfur dioxide—a heavy, colorless gas of pungent odor. It is a chemical benefactor as a bleaching agent and disinfectant, and the parent substance, in a sense, of sulfuric acid, one of the most widely and variously used of the products of industry. The great bodies of ores of copper, iron, zinc, and lead are sulfides, which, when the ore is exposed to high temperatures in the presence of a sufficient amount of air, burn to sulfur dioxide. If the supply of air is restricted, some of the sulfur may escape as such and be condensed in the flues or in the outer air as elemental sulfur. On the other hand, if there is a liberal excess of air, especially with certain ores, some of the sulfur may be converted to that final product of its oxidation, sulfur trioxide. This, too, is a gas at elevated temperature but a solid at ordinary temperatures, with remarkable affinity for water, with which it combines to form sulfuric acid. One of the older ideas which it has been most difficult to dislodge was that this substance, into which sulfur dioxide was supposed slowly to be converted in the air or on the foliage of plants, was the real agent of destruction with which we have to deal in smelting districts. This, however, is certainly not the case. Plants will, in fact, withstand several hundred times the concentration of sulfur trioxide that they will of sulfur dioxide. It is plainly visible in clear atmospheres carrying it even to the extent of one part in twenty million parts of air.

It is about its near relative, sulfur dioxide, that most of the smoke litigation of this and other countries has centered, and like most offenders before the law, its bad reputation, unfortunately, has often gone far toward establishing its guilt. In the reckless days of heap roasting when smoldering piles of ores rich in sulfur were burned in open bins, clouds of gaseous and finely divided solid products were produced, chilled quickly by contact with the cooler air, and swept by ground currents across country practically at the level of vegetation. The result is all too familiar to the older residents of the great smelting districts of the South and West. Extensive areas within the radius over which the gases traveled were denuded of plant life before the sulfur dioxide they carried was reduced to the point where it was noninjurious through its absorption by the soil, by vegetation, or through its dilution by the air.

Then came that period of rapid industrial expansion of fifteen or twenty years ago in which the genius of America for developing large industrial enterprises found notable expression. Out of this came a number of great smelters, one of which alone treated daily upwards of ten thousand tons of ore, produced one-sixth of the world's supply of copper, and discharged daily through a single tall stack several thousand tons of sulfur dioxide gas, among other waste products. The development of the modern smelter put an end to heap roasting, but not in every case to sulfur dioxide injury. In many instances trailing smoke streams from tall chimneys reached the ground farther away, and, in less concentrated form, slowly extended to much wider limits the area of appreciable injury. In many districts this brought within a few years so many new complainants that powerful agrarian interests were pooled to wage war by injunction against the offending plant. The consequence was a situation injurious to all parties concerned. When damage was done, there was no reasonable basis for its adjustment. Few men were skilful enough by training and experience to distinguish sulfur dioxide injury to plants from many agents—as drouth, frost, insect pests, fungus or bacterial disease, or poor farming—which might easily simulate it. The result was a tendency, on one side, to ascribe to these other factors all cases of sulfur dioxide injury, and, on the other side, to assign to the well-known fact that a smokestack in the neighborhood was belching forth a gas which had an injurious effect on vegetation, all the season's disappointments in crop production.

A very important element in the general situation was the belief that repeated mild fumigations with sulfur dioxide in concentrations insufficient to show typical foliar markings were nevertheless injurious, contributing ultimately to a reduced vitality, to a less flourishing condition of the plant, and consequently to a reduced crop yield. This belief in a condition of "invisible injury" at once became a matter of overshadowing importance in smelter-smoke litigation, as did the question of "normal arsenic" in the human body in medico-legal cases following the work of Gautier on that subject. Manifestly, such a theory was a menacing one, for, if it were sound, any industrial operation which contributed sulfur dioxide to the atmosphere was not merely a potential, but an actual agent of injury. It was thus a subject of injunctive relief as a nuisance in the eyes of the law.

There is, in fact, no injury which is visible in this sense. Exposure to sulfur dioxide leads to no injurious effects unless it reaches concentration which produce characteristic foliar markings. Extended low-concentration fumigation experiments carried on daily through the entire growing season with various plants have afforded conclusive evidence on this point. Plants thus treated have shown a higher sulfur content and a higher protein content than the controls, but no reduction in crop yield. The carbohydrate and fat content was not altered. There is evidently a concentration of sulfur dioxide for each species of plant which represents a threshold of tolerance below which it is absorbed without injurious effects, and may even be utilized directly as a food.

Method for Estimation of Sulfur Dioxide in Air

A new epoch in the smelter-smoke controversy began when Marston and Wells developed a remarkably accurate and rapid method for the estimation of sulfur dioxide in air. In this method for the determination of sulfur dioxide a blue solution is made by adding a small amount of iodine solution to a solution of starch. This solution is divided into two equal portions, and each is added to one of two large tubulated glass bottles of about 20 liters capacity. One of the bottles is then aspirated to remove a certain amount of the air. A sample of the air to be tested is then admitted by opening a stopcock, the bottle meantime being vigorously shaken to insure complete contact of the air with the blue solution. To make all conditions equal, the other bottle, used as a standard, is shaken simultaneously for the same time in the same way. The two blue solutions are then removed to small bottles, and the colors compared. If sulfur dioxide was present in the air sampled, the test solution will have a lighter blue color. Its color is then restored exactly to that of the standard by adding a very accurately measured amount of a standard solution of iodine, which amount becomes directly a measure of the amount of iodine reduced by the sulfur dioxide, and thus indirectly an exact measure of the latter in the aspirated air.

Nothing more need be said of the accuracy of this method than that, in skilful hands, and within the concentrations of sulfur dioxide found in the air of smelting districts, the probable error is not over one part of sulfur dioxide in ten million parts of air, by volume. The method is not only one of great accuracy, but it is so rapid that, without undue haste, a complete determination can be made in less than five minutes. It is a common practice, in studying smoke-stream concentrations, to make hourly "runs," made up of a determination every five minutes. In the field specially equipped automobiles could follow a smoke stream and, in the course of one season of plant growth from springtime to the autumnal blighting, could pile up thousands of accurate determinations of sulfur dioxide at the level of vegetation at various distances from a smelter under known conditions of wind, humidity, and other weather conditions. At experimental farms, hundreds of plots representing all the common crops of the district could be subjected to fumigation by moving volumes of air containing accurately known concentrations of sulfur dioxide, under known conditions of light, temperature, duration of exposure, and maturity of plant.

The value of this method in the scientific investigation of sulfur dioxide as an agent of injury cannot be estimated. At once, through it, a quantitative basis was laid for a systematic attack in the open field as well as the experimental fumigation plot on the whole problem; and out of this grew two investigations of fundamental interest and importance—that of the Selby Smoke Commission in California, who were the first to employ carefully controlled moving atmospheres containing known amounts of sulfur dioxide in artificial fumigation experiments; and the later extensive investigation carried on under the direction of P. J. O'Gara, of Salt Lake City. The latter is a scientific investigation of outstanding merit, and represents the greatest single contribution yet made to the study of sulfur dioxide.

Effects of Sulfur Dioxide on Vegetation

There are many extremely interesting aspects to this question of sulfur dioxide injury to vegetation. In the first place, it is not the roots, nor the stems, but the green foliage which suffers. A plant not in leaf, as a deciduous tree in the dormant winter period, is immune from injury by atmospheric sulfur dioxide. The leaf is covered with an epidermis which is practically impervious to moisture and air, but over its surface are scattered many breathing pores, or stomata, which are the sole paths of entry or exit for absorbed or eliminated gases. At the mouth of each stomate are two large guard cells, like portly wardens, which by swelling or contracting control the stomate opening. During the hours of darkness when metabolic processes in plants are at a low ebb, or during periods of very low temperature, or of dry and hot winds when the loss of water may be excessive, the opening is contracted. Access to the leaf through the stomate is restricted, and, as would be expected, the plant at once becomes more resistant to injurious gases. There is here precisely the explanation of the fact that the concentration of sulfur dioxide in air which will produce foliar markings in daylight must be increased many times to do so at night. It has often been claimed that the greatest damage to vegetation occurs in the very early morning hours before sunrise, while the dew is on the grass, or during rainy weather when sulfur dioxide, being soluble in water, would be dissolved out of the air in smelting districts and thus reach the plant in concentrated form. It is true that this gas is very soluble in water—about eighty volumes of it dissolving in one volume of water, at standard conditions of temperature and pressure—but this solubility varies with the partial pressure of the gas in any gaseous mixture. In air containing ten parts of sulfur dioxide per million parts of air, which is a high value for the smoke stream of a smelter at the level of vegetation, the partial pressure of the gas would be only one hundred thousandth of an atmosphere, and its solubility would be only one hundred thousandth of this amount. Not the slightest injury could possibly result from spraying plants with such dilute solutions of sulfur dioxide. In fact, at atmospheric temperature and at sea level, water through which air containing ten thousand parts of sulfur dioxide per million has been passed will not cause injury.

Another factor is temperature. A plant is much more resistant to sulfur dioxide at temperatures of 5° C. and below than it is at higher temperatures; yet, there is no considerable variation above 5° C. within the usual range of atmospheric temperatures.

Duration of exposure is a factor of great importance. One of the notable results of Dr. O'Gara's work, which has covered about two hundred fifty species or varieties of plants, has been the formulation of a law of gas action on the plant cell in which this fact is clearly expressed. This law states that, with other environmental factors constant, the active part of a gas necessary to produce a certain effect on a plant cell, multiplied by the time through which it acts, is a constant. This law can be expressed by the fundamental equation

$$t(L - l) = C$$

in which t is the time through which the gas acts to produce a certain effect, L is the concentration of the gas in the atmosphere to which the plant is exposed, l is that concentration of the gas which will not act injuriously on the plant cell if applied for an infinitely long time, and C is a constant.[1] The equation may be changed to include environmental and other factors, such as light, relative humidity of the atmosphere, temperature, stage of growth, and kind of plant—all of which can be experimentally determined, and thus the whole question of plant injury by this agent can be placed on a quantitative basis. The concentration of gas necessary to cause injury can then be indicated thus:

$$L = R\left(\frac{C}{t} + l\right)$$

where R is a factor or multiple factor to correct for the above environmental and other conditions. It is, of course, multiple whenever more than one of these conditions acts simultaneously to increase the resistance of the plant to the gas.

[1] This Journal, 14 (1922), 744.

The importance of the time factor in smoke injury indicates at once that a region of variable winds will be much less subject to injury than will one of steady winds. The gas stream is whipped back and forth across a broad sector, dilution is promoted, and damage to vegetation, in which the time factor of continuous exposure is a vital one, may be completely prevented, even when with a steady wind, prevalent as to direction, serious injury might result. Strong winds on land are often unsteady in direction near the ground. It is a common sight to see a lofty weather-vane ranging in its direction at intervals over a considerable arc during windy periods. Stack gases can reach distant vegetation in any other than highly dilute form only as they are carried there by air currents, and only if these currents are steady can a long fumigation of a given plant result; and as these winds are strong or gentle will the time required to carry the stack gases there be short or long, and the concentration when they do reach the plant be relatively large or small. Wherever hot gases are discharged through tall stacks in open level country it is almost impossible on still days for them to reach the level of vegetation before they are so extensively diluted that no injury could result.

It is remarkable, but true, that the relative humidity of the atmosphere is one of the most important factors in controlling the effect of sulfur dioxide on vegetation. A plant will endure six or eight times the concentration of sulfur dioxide at 50 per cent relative humidity that it will in a humid or moist atmosphere. This generally means that, as during the hours of darkness, so during days of very low humidity, the plant life of a district would be immune to injury. There is not much increase in sensitiveness from 70 per cent relative humidity to 100 per cent, but the resistance of a plant increases more rapidly when this drops below 70 per cent, and becomes very prominent below 50 per cent.

It would appear, then, that in those smelting districts in which the concentrations of sulfur dioxide are uniformly low, yet sufficient at times to cause a "burn," as it is often termed, on vegetation, four conditions must be coincident for damage to occur. These are daylight, an atmospheric humidity of 60 per cent or over (allowing a margin here), a temperature of above 5° C., and a prevailing wind, steady as to direction. This scientific prescription was once proposed and followed for one season as a basis for operation for one of the large western lead smelters. An observer went on duty at daybreak at the plant weather station and from then until sundown watched carefully for the simultaneous onset of the conditions mentioned above. During the season of plant growth the temperature is usually above 5° C., so this factor was usually prevalent. If in addition there was a humidity of 60 per cent or over and the wind blew in one direction for an hour, a critical condition was approaching and orders were issued to shut down roasting operations at the plant—this being the chief source of the sulfur dioxide. Thus, with atmospheric conditions most favorable for sulfur dioxide injury, the amount of sulfur dioxide in the smoke stream was promptly reduced to such a low level that no damage could result. Operation was maintained on this basis until the humidity decreased, the wind direction changed, or darkness came on, when normal operations were restored. Manifestly, such a program could only be adopted in regions subject to variable winds and to long periods of low atmospheric humidity. Even then it means occasional sharp interruptions in roasting operations, and this is objectionable. At the same time, however, it stands as an interesting application of the results of scientific investigation to the temporary relief of a threatening situation. This plan of operation has been known as the "Sea Captain Theory," from the good-natured protest of the plant manager when it was proposed to him, "What, do you expect me to get a sea captain to run my plant?"

Finally, different kinds of vegetation vary greatly in their relative resistance to sulfur dioxide. Under field conditions, barley, alfalfa, and oats rank among the most sensitive of the common cultivated plants, and will show foliar markings if treated for three hours, under optimum conditions of light, temperature, and humidity, with air containing one part of sulfur dioxide per million parts of air by volume. Wheat and beets require somewhat more. The white potato is more resistant, with about three parts per million required. Corn would show foliar markings at about four parts, and celery at a little over six. This varying susceptibility of different species of plants is often of valuable assistance in the field in identifying sulfur dioxide injury, for if signs of injury have been observed on one kind of vegetation and the relative resistance of this is known, one should expect to find similar effects on any neighboring plants of known equal or less resistance. Sulfur dioxide injury, when it occurs, is quite characteristic for a given plant, usually taking the form of elongated bleached areas between the veins or along the margin of the leaf. Where it is severe and recent the veins often stand out prominently and green between these bleached areas. It generally requires from one to three days for a "burn" to develop, the first indications usually being the appearance of faded green areas which slowly change to pale brown, yellow, or white. The question of how far a "burn" on a given crop during its growth will express itself in the crop yield is an important one. Does the injury amount merely to a certain local destruction of leaf tissue, or are there other more general effects?

Slight markings repeatedly produced on plants during the season of growth evidently cause no appreciable loss in crop yield, and in the earlier stages of growth plants may suffer a considerable loss of leaf tissue by sulfur dioxide injury and recover completely with no measurable loss in crop yield. The most notable effects on the crop are found near the end of the growing season, when injured leaf tissue is not so readily compensated for by the production of new leaves.

Absorption from Air

Sulfur dioxide gas is rapidly absorbed from the air, both by the soil and by vegetation. In fumigation experiments one is astonished at the rate at which this gas disappears from still atmospheres. In such cases the exposure of plants to an atmosphere of constant composition, as far as sulfur dioxide is concerned, is impossible, and precisely on this account, if on no other, the older work on this substance as an agent of injury to plant life, in which certain amounts of it were added to closed fumigation chambers, is of no value whatsoever when one attempts to express it in terms of prevailing field conditions. Even where moving streams of air containing known amounts of sulfur dioxide were sent horizontally through long fumigation cabinets, it was observed that the air emerged with much less sulfur dioxide than it had when it entered. We have here an explanation of the fact that, remote from sources of this gas, the air carries no appreciable amount of it. There is no evidence pointing to any harmful effects to the soil from exposure to sulfur dioxide. Experimental plots which have been exposed daily to fumigation over long periods with such concentrations as are found in smelting districts at the present time, have shown no loss in fertility in producing subsequent crops.

Solution of Problem

The road to an exact appraisal of field conditions in any smelting district is now open. There are, of course, other important features besides those mentioned above which give a certain individuality to every smelting district—the topography of the country, the kind of vegetation, the climate, the character of ore smelted, and the methods pursued in its treatment—all of which must be considered before any remedy can be offered to relieve a critical condition. But the key to the whole matter lies in a knowledge of prevailing sulfur dioxide concentrations near the surface of the ground in every sector of the zone of probable injury. If these are too high they must be reduced. This may be accomplished either by the removal of sulfur dioxide at the source by converting it to sulfuric acid, or by increasing its dilution with air before it reaches the level of vegetation, to a point where injury will not result under conditions most favorable for its action, on the most sensitive plants in the district. Unfortunately, the first method is necessarily of limited application. Relatively few of the smelters of this country can produce, from the ores treated, gases rich enough in sulfur dioxide to convert this economically to sulfuric acid, and, once produced, find a market for it.

A notable achievement in this connection is that of the smelter in Tennessee, which treats a low-grade copper ore containing about 27 per cent of sulfur. With the installation of newer methods about twenty years ago, a tall stack was erected and from this were discharged all the waste products of smelting operations. Serious complaints of injury soon arose across the Georgia border, far to the south of the old denuded but limited area over which heap roasting had left its permanent effects. Injunctive relief was sought and secured by the state of Georgia on behalf of the complainants, and at once steps were taken by the management to find a way out of the difficulty by converting the sulfur dioxide to sulfuric acid. This now stands as one of the great industrial achievements of this country. Baffling obstacles arose on every hand. The enormous volumes of furnace gases to be treated, their high temperature, fluctuating composition, the presence of carbon dioxide as a new factor to be dealt with—all conspired to produce a situation which required much painstaking experimental work and a courageous management. The outcome was the largest sulfuric acid plant in existence, producing in 1920 over 330,000 tons of acid, most of which is now used to convert Florida phosphate rock to acid phosphate fertilizer. Thus did an injurious waste product of industry become a major product, the chief source of revenue to the plant which produced it, and at the same time an agent dedicated to the purpose of increasing the fertility of the plantations of the South.

Another case of particular interest is that of a smelter in Montana, which for twenty years has represented the greatest achievement in metallurgical practice, not only in the amount of ore treated daily, which has grown to sixteen thousand tons, in the copper produced, which is approximately one-third of the total production of this country, but also in the progressive methods employed in plant operation. Its huge smoke stream, discharged from a tall stack on a spur of the Rocky Mountains, was a conspicuous object at all times, and contained two substances in particular which were a menace to the surrounding territory. It carried several thousand tons of sulfur dioxide daily at a dangerous concentration, and in addition a large amount of suspended solids, the chief component of which, amounting to twenty tons or more per day, was arsenic trioxide, a substance poisonous to animal life. As a consequence, a zone of injury began creeping slowly year by year particularly southward into the mountain forests, and arsenic in dangerous amounts was distributed on vegetation over a wide area. Both of these agents of possible injury are being relieved—on the one hand, through the removal of some of the sulfur dioxide in the manufacture of sulfuric acid for use in making acid phosphate from great near-lying deposits, and the discharge of the rest through a very high stack; and on the other hand, through the more complete recovery of the solid waste products by settling chambers and by the Cottrell process of electrical precipitation. Here again, a waste product is becoming a by-product indispensable to agriculture as an insecticide. The use of arsenic, particularly in the form of arsenates of lead and of calcium, is increasing so rapidly that even the large output from smelting operations in this country is unable to meet the present demands.

Where conversion to sulfuric acid is an impossible remedy, the tall stack and a high temperature in the exit gases provide the best probable means of obtaining the necessary dilution of the sulfur dioxide.

The most conspicuous examples of the application of these methods to the solution of smoke difficulties of long duration are those of neighboring lead smelters at Murray and Midvale, Utah. Here in a fertile valley of surpassing beauty of location, just south of Salt Lake City, two smelters are located in the center of a highly cultivated and prosperous region of varied and sensitive crops. For many years they were involved in litigation which threatened permanent injunction against further operation, until work to which reference has already been made pointed the way to a safe plan of operation. This has involved the discharge of the gases from roasting operations from a tall stack, and the maintenance, during the growing season for plants in that district, and during the hours of daylight, of a temperature difference of at least 75° C. between the stack gases and the outer air. The gases are preheated to the required extra amount during these periods by coal burners near the base of the stack.

There has been brought here to the recognition it deserves a very important but neglected factor in smoke control. The question of whether stack gases are heavier than air and will tend to fall, or are lighter than air and will rise to higher levels, is much more a matter of temperature than of composition. The oft-repeated claim that stack gases from smelting operations are heavy and fall rapidly to the ground is usually fallacious, for, while it is true that they may, and usually do, have a specific gravity greater than that of air, one has only to express this in terms of figures to see how insignificant is the difference and how slight is the change in temperature which will completely overcome it. Air containing 1 per cent sulfur dioxide is heavier than pure air, but if it is 3.5° C. warmer than air it will be lighter. Anyone who has seen a toy balloon filled with hot air go bounding skyward has an impression of what occurs when hot gases, in such huge columns as emerge from the mighty smoke stacks of large smelters, are discharged into the air. How tall shall be the smoke stack or how high shall the exit stack gases be heated are questions that must be settled for the individual smelter, but in general it may reasonably be contended that for every smelter operating along modern lines there is a combination of high stack and high temperature of exit gases which will provide a remedy for injury by sulfur dioxide.

Notable advances have been made in the recovery of solid waste products of smelting operations. The high temperatures and powerful drafts lead necessarily to the volatilization or mechanical sweeping up of metallic compounds which were formerly carried out through short flue systems and low stacks and distributed over the outlying country. The extent of these losses is indicated by the fact that it was found very profitable sometime ago to scrape the top soil from several square miles of land surrounding one of the old Montana smelters, concentrate it, and smelt the residue for copper and other metal values it contained. An interesting development, but of limited application, is bag filtration, in which the enormous volumes of furnace gases are treated with lime to neutralize free sulfuric acid and sent through woolen bags, vertically hung, open mouth down, in sectional bag houses. The removal of solids is practically complete, so that not even a haze marks the path of the gases issuing from the exit stack. Wherever it has been employed this process has settled at once all question of damage from solid emanations and brought good returns from the large amount of dust recovered. Unhappily, it has often increased sulfur dioxide damage simply because bag filtration requires precooling of the furnace gases to 90° C. or less, and the gases finally pass out the stack so cold that they possess little buoyancy. Of much wider application is the Cottrell system of electrical precipitation of solids and mists from smelter smoke. The furnace gases need be cooled only to the point where condensation of volatile solids will be fairly complete, and the neutralization of sulfuric acid is not required. In fact, the process is aided by the presence of this substance. In the Cottrell process, particularly, because of its wide application, we have an admirable conservation measure, as well as a process which is of growing importance in relieving the smoke evil. Its use is being extended in many directions, notably in dust recovery from cement plants, abnormal local dust fall, and litigation in the West where a dry raw-mix, and oil-fired rotary kilns are widely employed. The dust loss from these kilns will often range between five and seven tons of dust per day or from fifty to seventy tons from a ten-kiln plant. With proper installation and control the Cottrell process will recover upwards of 95 per cent of this loss. This dust often carries soluble potassium salts in quantities which have often justified their recovery as a by-product of the plant.

LEGAL ASPECTS

The courts of this country are holding almost unanimously to the principle of granting unqualified protection to established property rights, without respect to the magnitude of the interests involved in the litigation. The fact that a great capital investment lies behind an offending industrial plant with a payroll so large that it has a direct bearing on the prosperity of the whole community, is not acceptable as a legal defense, nor even a mitigating circumstance to justify any invasion of the right to the enjoyment of property. This assumes, of course, that great interests should not be overthrown on trifling or frivolous grounds. But while it is conceded that every property owner has a right to devote his own property to any legitimate pursuit, this right is restrained to the extent that there shall be no substantial interference by him with the comfortable enjoyment of his neighbor's home. In other words, there can be no balancing of conveniences when this involves the preservation of an established right.

The injury which constitutes a nuisance must be a real one, such as impairs the enjoyment physically of the property within its sphere. If it be slight or trivial or fanciful, or one of mere fastidiousness, there is no nuisance in a legal sense. Thus, the law will not declare a thing a nuisance because it is unsightly or disfigured or unpleasing to the eye, or a violation of good taste, for the law does not cater to the tastes of men or consult their convenience, but only seeks to uphold their material rights, and to protect them in the ordinary comforts of human existence. Nor can the trespasser say to the owner, "Your land is not adapted to the crops you grow; it is underlaid with hardpan, frequented by frosts, and swept by the hot winds. It is of little actual value; therefore, you are not seriously injured if I do destroy what little you have."

Agriculture now is, and will in all probability continue to be, the fundamental source of food supply of the world. As such it at once becomes the most important industry of the world. Scientific efforts to make our tillable lands more productive will not be relaxed in the future. The principle recently adopted by the Colorado River Commission, in which a liberal allotment is granted agrarian interests before the industrial development of power is allowed, implies a recognition of this same fact. Thus, the principle of the greatest good to the greatest number locally has rarely been accepted in law, and on this account the invasion of an agricultural community by an industrial plant, even of great magnitude, is dangerous to the industry if in the course of its operation it does injury to its neighbors. On the other hand, where agrarian and industrial interests can live together in peace, both contribute to the welfare of each. The isolated industry far from the perishable products of the farm and from an agricultural community is crippled by unfavorable and expensive living conditions for its employees. And the isolated agricultural region has only remote markets for its produce. The development of both side by side is thus equally important to both.

This is the direction in which the age-old issue of smelter-smoke injury is moving in this country. There are very few, if any, outstanding cases of progressive economic injury that are yet unsettled or not on the road to settlement. This has come about not always through a sheer desire to avoid doing injury to adjacent interests. A more important factor has been a better control of plant operation, more attention to stack losses, new markets for by-products, and less of single-mindedness in management. The old period which refused to recognize the interests either of the community or of the nation could not endure. The age of by-products has begun, tardily but truly, in America, and this means not only industrial independence, but national progress in the years to come.

Chandler Medal Award
Presentation by George B. Pegram

The Charles Frederick Chandler Foundation of Columbia University was established in 1910 by friends of Professor Chandler to provide from time to time a medal to be presented to an eminent chemist in recognition of his achievements in science, and to provide also for a lecture by the medalist. Surely everyone in this audience, after hearing the address of Professor Swain and being led by him to visualize the economic and social significance of progress by economic studies in connection with one aspect of modern industry, will feel that the Trustees of Columbia University were well advised by the Committee on the Chandler Medal when they made this, the seventh award of the medal, to Robert Eckles Swain, alumnus of Leland Stanford and Yale Universities, former student at Strasburg and Heidelberg, professor of chemistry and executive head of the department of chemistry in Leland Stanford University, who began his scientific career as a worker in physical and theoretical chemistry, was led on into notable applications of chemical knowledge in physiological studies, and from these to the broader field of the economic aspects of chemical industry, in which field he has organized and directed immense researches.

Professor Swain, for your distinction as a teacher, as an organizer of instruction and research, as a chemist of high accomplishment and of broad public outlook, this medal of gold is presented to you. But you will the more appreciate it to receive it not from my hand but from the hand of him whose name is given to it and whose likeness the artist has so finely molded upon it. Professor Chandler, will you kindly convey the medal to Dr. Swain?

Professor Chandler was escorted to the platform and placed the medal in the hands of Professor Swain with appropriate remarks as to the great importance of such work as Dr. Swain had described in his address in the salvage of waste products and the prevention of atmospheric pollution.

Bei Fragen zur Produktsicherheit wenden Sie sich bitte an:
If you have any questions regarding product safety,
please contact:

Walter de Gruyter GmbH
Genthiner Straße 13
10785 Berlin
productsafety@degruyterbrill.com